Machine Learning

A Comprehensive, Step-by-Step Guide to Learning and Understanding Machine Learning Concepts, Technology and Principles for Beginners

TABLE OF CONTENTS

Introduction

I want to thank you for choosing this book '*Machine learning - A Comprehensive, Step-by-Step Guide to Learning and Understanding Machine Learning Concepts, Technology and Principles for Beginners.*'

Machines are used in most households and their capabilities have evolved beyond performing manual tasks. There are some countries that use machines in their armies and some companies are using machines to perform some menial tasks. Now, machines can also work on tasks that require some cognition. Human beings were the only ones who had the ability to perform these tasks in the past. Predicting the outcome of tournaments, playing chess, driving cars, diagnosing diseases are some examples of the complex tasks that machines can perform.

However, the remarkable capabilities of machines have instilled fear in some people. They are wary of how powerful machines are and how they can change the world. For example, if you have watched Doctor Who, there are some robots, called Daleks, who wanted to take over the Earth because they were smarter than human beings. Skeptics fear that they may lose their jobs and some fear that the world may be taken over by robots and machines because they are smarter than human beings. The former is a valid fear since there is a possibility that machines can perform your job better than you. The BBC conducted the "Will robots take over my job?" survey and concluded that jobs like taxi drivers, actuaries, accountants, bar workers and receptionists will be automated soon.

The latter fear is not valid since it is difficult to teach a machine. Learning is a process that includes many smaller processes. In this book, we will look at the different ways a machine can learn. You will notice that the processes used to teach machines are similar to how human beings learn.

Research on automation must be read with a slight level of skepticism since the future of machines and artificial intelligence is unknown. Technology is moving fast, but its adoption is an unchartered path with unforeseen challenges. Machine learning is not simple since it does not only involve turning switches on and off. Machine learning is also not an out–of–the–box solution. Machines operate in parallel to the statistical algorithms, which machine learning engineers and data scientists often oversee. Industry experts believe that there could be a time when there will be an inadequate supply of people to operate machines.

Your path to becoming an expert in machine learning probably starts from here. But, you can also satisfy your thirst for knowledge with a base understanding of what machine learning is for now. You do not need to make hasty decisions. Let us proceed with the assumption that you want to train to become a machine learning engineer or data scientist. This book will help you achieve either goal.

If you are a beginner, this book will act as a guide. It covers all the information needed to understand machine learning better. In the first few chapters of the book, you will gather information on what machine learning is and types of machine learning. It also covers information on different algorithms that engineers developed to improve machine learning. Experts recommend that you practice and build projects to understand machine learning better. This book leaves you with some projects and ideas that you can use to enhance your learning.

Thank you for purchasing the book. I hope you gather all the information you are looking for.

Chapter One

What is Machine Learning?

Machines with artificial intelligence often learn processes and identify problems and solutions through machine learning. Some uses of such machines are data analysis, recognition, predictions, projections and diagnosis. They learn from training data, a sample dataset that resembles the complete population, to identify different patterns within the data and use those patterns to find a solution to a variety of problems. There are different learning mechanisms used to help a machine learn, of which supervised learning, unsupervised learning and reinforcement learning are the most common. We will cover these concepts in the following chapters.

In simpler words, a change made to the structure of the machine to enhance the performance of the machine is a type of machine learning. It is important to remember that not every change made to a machine helps the machine learn. For example, if a machine must predict which team will win the Barclays Premier League, the programmer will train the machine with historical information about the team's performance and the players' performances. Using this information, the machine can identify the patterns and correlations within the dataset. It can then use those patterns and predict which team will win the Barclay's Premier League.

Skeptics often wonder why machines must learn since they are wary of what these machines are capable of. Facebook had used machine learning and artificial intelligence to develop a machine trained with a specific vocabulary dataset. However, this machine began to develop its own language using the alphabet in the dataset. If multiple machines learned that language they can communicate with each other thereby making it difficult for us to understand what they are saying. Regardless of what the outcome was, Facebook and Google are trying to help machines learn through training datasets.

There are many reasons why machine learning is important. Psychologists are using some concepts of machine learning to help them understand human learning better. Machine learning also helps to improve the accuracy and efficiency of machines. It also helps to reduce the volume of code that a programmer writes to help the machine learn. There are times when a programmer may forget a small part of the code, which can lead to constant errors. It takes time for the programmer to identify this error in the lengthy code. Machine learning reduces the volume of code that a programmer writes thereby reducing the effects of human error.

Subjects involved in machine learning

Machine learning uses concepts from different subjects. The subjects mentioned in this section are not an exhaustive list since the field of machine learning is developing as you read this book. However, the concepts in the subjects mentioned below, especially those in statistics, are the foundation of machine learning.

Statistics

The concepts in statistics like regression, clustering, data analysis and hypothesis testing are the foundation of machine learning. Most machine learning algorithms use these concepts to train the machine. Training is a common problem in both statistics and machine learning. Training is the process where different sample datasets help the machine draw information about the population. The machine stores the information and uses it to predict or project futuristic values. The machine can also identify a problem and a solution to that problem using the training dataset. Another problem that is common to both statistics and machine learning is the identification of the values of a function at a given point. Solutions to such problems are instances of machine learning since problems that involve the estimation of future events often use data about past events.

Brain modeling

A concept of brain modeling called neural networks is closely related to machine learning. Scientists believe that one can replicate the model of the neurons in the human brain and build a non-linear neural network. The layers in this network have nodes and the engineer can assign weights to those nodes to calculate the output. Psychologists and scientists are now using neural networks to understand human learning better. Sub-symbolic processing, connectionism and brain style computations are a few spheres that they are exploring to understand human and machine learning better.

Adaptive control theory

Control theory studies how systems control their efficiency by adapting to a change in the system. A problem that most systems face is the change in the environment. A system uses different methods to adapt to the changes in the environment and continue to perform effectively. The idea is to make the system capable enough to anticipate any changes in the environment and adapt to maintain the efficiency and accuracy of the machine.

Evolutionary models

Evolutionary studies conclude that human beings and animals both learn to not only adapt to a change in their environment but also learn to perform better. The concepts of learning and evolution can are the same for machine learning and human learning. The methods that psychologists use to understand human learning help scientists improve how a machine learns.

Varieties of Machine Learning

So far, this book has introduced machine learning and has answered the question about the subjects that constitute it. Now, we come to the more important question of what can be learned about machine learning. The following are a few topics on which knowledge can be gained through the study of machine learning:

- Programs and logic rule sets

- Terminology and grammars

- Finite state machines

- Problem - solving systems

- Functions

- Artificial Intelligence

- Statistics

Out of the above, the two most focused on topics are those of statistics and artificial intelligence. These two subjects are used extensively in machine learning. We now move on to chapters that describe the two broad categories of machine learning techniques: supervised machine learning and unsupervised machine learning.

Uses of Machine Learning

Most organizations and enterprises use machines to complete manual tasks that were once impossible to complete in a short time, especially if large volumes of data were used. Over the last few decades, the volume of data available has rapidly increased making it impossible for human beings to analyze that data. This increase in data paved the way for automated processes and machines to work on tasks that are difficult for human beings to complete.

One can derive useful information from data analysis and this information can help us drive our businesses and lives. We have set foot in the world of Business Analytics, Big Data, Data Science and Artificial Intelligence. Predictive analytics and Business Intelligence are no longer just for the elite but also for small companies and businesses. This has given these small businesses a chance to participate in the process of collecting and utilizing information effectively.

Let us now look at some technical uses of machine learning and see how these uses can be applied to real-world problems.

Data replication

Machine learning allows the model or the algorithm to develop data that looks like the training dataset. For example, if you run the manuscript of a book through a machine, the machine identifies the density of the words in every page of the manuscript. The output is text that is similar to the information in the manuscript.

Clustering variables

Machines use the clustering algorithm, covered in a later chapter, to group variables that may be related to one another. This tool is useful when you are unaware of the how a change in the variables affects the machine. You can also gather information on how a change in one variable affects the other variables in the group. When the dataset is large, the machine can look for latent variables and use them to understand the obtained data.

Reduction of Dimensionality

Data always has variables and dimensions. It is difficult for human beings to analyze data that has more than three dimensions. Machine learning helps to reduce the dimensions of the data by clustering similar variables together. This process helps a human being understand and identify relationships within and between the variable clusters.

Visualization

There are times when the user may want to visualize the relationship that exists between variables or obtain the summary of the data in a visual form. Machine learning assists in both these processes by summarizing the data for the user using specified or non – specified parameters.

Chapter Two

Facts about Machine Learning

Machine learning is permeating numerous aspects of our everyday lives, right from optimizing Netflix recommendations to Google searches. Machine learning has contributed to improving different facets of building mechanics in smart building space and the experiences of the occupant. You do not have to have a Ph.D. to understand the different facets and functions of machine learning. This section covers some facts about machine learning that are very basic and important to know.

Bifurcation of Machine Learning

Supervised and unsupervised machine learning are two techniques that programmers and scientists use to help machines learn. Smart buildings incorporate both types. Here is a simple example of how these types of machine learning look like: Let us assume that you want to teach a computer to recognize an ant. When you use a supervised approach, you will tell the computer that an ant is an insect that could either be small or big. You will also need to tell the computer that the ant could either be red or black. When you use an unsupervised approach, you will need to show the computer different animal groups and then tell the computer what an ant looks like and then show the computer another set of pictures and ask the computer to identify the ant until the computer learns the features specific to an ant.

Smart building spaces use both supervised and unsupervised machine learning techniques. The applications in these smart buildings allow the users to provide feedback to the building to improve the efficiency of the building.

Machines are not fully automatic

Machine learning helps computers automate, anticipate and evolve but that does not mean that they can take over the world. Machine learning uses algorithms that human beings develop. Therefore, machine learning still needs human beings since they will need to set parameters and train the machine with different training datasets.

Machine learning helps a computer discover patterns that are not possible for human beings to see. The computer will then make an adjustment to the system. However, it is not good to identify and understand why those patterns exist. For instance, smart buildings human beings created smart buildings to ensure that the people inside the building help to improve the living conditions of the people. However, one cannot expect that a machine will learn to become more productive. A human must set up the definitions and rules that the building will need to follow.

It is important to note that the data cannot always explain why any anomalies or outliers occur. Consider a scenario where the people in the building constantly request that the temperature of the building must reduce by 20 degrees when compared to the external environment. The machine-learning algorithm will take this request into account and notify the operator about the request. Therefore, it is important for skilled people to operate machines to ensure that the conclusions obtained are accurate.

Anyone can use machine learning

Writing a machine-learning algorithm is very different from learning how to use that algorithm. After all, you do not need to learn how to program when you use an app on your phone. The best platforms always create an abstract of the program to present the users with an interface, which need minimal training to use. If you do know the basic concepts of machine learning, you are ready to go! Data scientists must edit or change the algorithms.

Machine learning has come of this age and is growing quickly. Buildings are using machine learning in different ways to make the existing infrastructure efficient and help to enhance the experience of the occupants residing in the building. Right from an energy usage standpoint, buildings are always learning and analyzing the needs of the occupants.

How does this affect us going forward? This advance in machine learning goes to say that most things will happen without the need for us to ask. Machine learning engineering could go beyond managing lighting and temperature. Machine learning implies that there will be some future state of multiple layers and levels of automation adjusting based on the current activity.

Data Transformation is where the work lies

When you read through the different techniques of machine learning, you will probably assume that machine learning is mostly about selecting the right algorithm and tuning that algorithm to function accurately. The reality is prosaic – a large chunk of your time goes into cleansing the data and then transforming that data into raw features that will define the relationship between your data.

Revolution of Machine Learning has begun

During the 1980s there was a rapid development and advancement in computing power and computers. This gave rise to enormous amount of fear and excitement around artificial intelligence, computers and machine learning which could help the world solves a variety of ailments – right from household drudgery to diseases. As artificial intelligence and machine learning developed as formal fields of study, turning these ideas and hopes into reality was more difficult to achieve and artificial intelligence retreated into the world of theory and fantasy. However, in the last decade, the advances in data storage and

11

computing have changed the game again. Machines are now able to work on tasks that once were difficult for them to learn.

Machine Learning and Artificial Intelligence are interrelated

Machine learning is a subset of artificial intelligence that drives the process of data mining. What is the difference between these terms? Experts spend hours debating on where they must draw the line between machine learning and artificial intelligence.

Artificial intelligence allows machines to think like human beings. At any given minute, human beings can capture thousands of data points using the five different senses. The brain can recall memories from the past, draw conclusions based on causes and effects and make decisions. Human beings learn to recognize patterns, but every being has a limit.

One can think of machine learning as a continuous and automated version of data mining. Machines use data mining to detect certain patterns in data sets that human beings will not be able to find. Machine learning is a process that can reduce the size of the data to detect and extrapolate patterns that will allow us to apply that information to identify new actions and solutions.

In smart building spaces, machine learning enables any building to run efficiently while also responding to occupants' changing needs. For instance, you can look at how a machine learning application can do more when compared to how a smart building may handle a recurring board meeting. However, any machine-learning algorithm can make sense of more than a thousand variables at any given time of the year to create an ideal thermal environment during the business meeting.

Chapter Three

Types of Machine Learning

Unsupervised and Reinforcement Machine Learning

Reinforcement learning is a technique that allows the machine to interact with the ambient environment through actions. The machine receives a reward if the environment reacts positively to the action. However, the machine will receive punishment if it receives a negative response. The machine learns from these responses and improves its performance to maximize a positive response from the environment. If you pay close attention to the description, you will relate reinforcement learning to human learning. For example, you feed a baby when she cries. The baby is aware that she can get food if she cries. She also knows that she can gather attention from people around her if she reacts positively.

Reinforcement learning is a technique that is related to decision theory in statistics and management sciences and control theory in engineering. The problems studied in these subjects are like the subjects that one studies in under machine learning. However, the subjects focus on different parts of the problem.

Unsupervised machine learning is another technique that uses some concepts from both reinforcement learning and game theory. However, in unsupervised machine learning, the environment is dynamic and can include many machines. These machines produce actions and receive rewards and the objective of every machine is to maximize the rewards. They must also assess how the other machines work. The application of game theory to such a situation with multiple, dynamic systems is a popular area of research.

Unlike reinforcement learning, a machine receives a training dataset if the engineer uses unsupervised machine learning to train the machine. However, the dataset does not contain information about the required output. This begs the question - how can the machine possibly learn anything without receiving any feedback from the environment or having information about target outputs? The idea is to allow the machine to develop input vectors that the machine can use to predict the output for different types of input datasets. Dimensionality reduction and clustering both use unsupervised machine learning techniques. The technique of unsupervised learning is closely related to the fields of information theory and statistics.

Supervised Machine Learning

Supervised machine learning uses training datasets to help machines learn. These datasets contain various examples that consist of the input and the desired output, commonly known as supervisory signals. Machines use supervisory learning algorithms that help to generate inferred functions that forecast or predict events. These functions are called classifiers if the output is discrete and regression functions if the outputs are continuous. The supervisory algorithm must conceive a generalized method that helps the machine reach the desired output. Human beings and animals learn in a similar way through concepts. For example, if you are in a trigonometry class, you will learn a set of functions that you should use to solve specific problems. Let us look at how the supervised machine algorithm works.

Step 1

The engineer must determine what types of example should be a part of the training dataset. The engineer must be careful with the training dataset since the machine learns from the examples in the set. He must ensure that the examples have the right input and output. For example, if the machine must learn to recognize speech, the engineer must provide examples of words, sentences, paragraphs or phonemes. If the engineer provides the machine with numbers or images, the machine is not going to produce the desired output.

Step 2

The engineer must collect the data and clean it before she uses it as the training dataset. The dataset must represent all the possible outcomes and functions that the machine can develop for some input. The engineer must ensure that she maps the input examples with their correct outputs.

Step 3

The engineer must now decide how the input examples she should provide to the machine. This is an important step since the accuracy of the machine is solely dependent on the representation of the input. Vectors represent the data and this vector contains information about the attributes and characteristics of the data. For example, if one of the inputs contains information on the products available at the store, the input vector can contain the following

- Product Name

- Brand

- Product Type

- Manufacturer

- Distributor and so on.

However, the machine must learn to only include some attributes to avoid a long training period. Many features may also lead to failure since the machine cannot identify or predict the output.

Step 4

The engineer must decide on the structure of the function that the machine should develop or use. She must also identify the algorithm that the machine should use to obtain the desired output. Common algorithms used are regression, clustering and decision trees.

Step 5

The engineer must then complete the design. She should run the algorithm on the dataset. There are some control parameters that the engineer must enter to ensure that the algorithm works well. Cross-validation methods help to estimate the parameters. If the dataset is large, it is best to break the set into smaller subsets and use the cross-validation method to estimate the parameters.

Step 6

Once the algorithms run and the machine generates the function, the engineer should measure the accuracy of the function. The engineer must use a testing dataset instead of the training dataset to check the accuracy of the function.

There are many supervised machine-learning algorithms in use and each of these algorithms has its strengths and weaknesses. The selection of the learning algorithm is an important step in the process since the engineer cannot use a definitive algorithm to train the machine.

Issues to consider in Supervised Learning

With the usage of supervised learning algorithms, there arise a few issues associated with it. Given below are four major issues:

Bias-variance tradeoff

The bias-variance tradeoff is an issue that every engineer should be wary of while working with machine learning. Consider a situation where we have various but equally good training sets. If the engineer uses different datasets to train the machine, the learning algorithm may form a bias to the input. Machines often give systematically incorrect outputs if the machine forms a bias to the input. Learning algorithms have a high input variance, which occurs when the algorithm causes the machine to predict different outputs for that input in each training set. The sum of the bias and variance of the learning algorithm is known as the prediction error for the classifier function. There exists a tradeoff between bias and variance. A requirement for learning algorithms with low bias is that they need to be flexible enough to accommodate all the data sets. However, if they are too flexible, the learning algorithms might end up giving varying outputs for each training set and increases the variance. Supervised learning methods need to be able to adjust this tradeoff that happens automatically. The alternative is to use an adjustable parameter.

Function complexity and amount of training data

The second issue is concerned with the training data. The training dataset must be prepared depending on what time of function the machine should generate, a classifier or a regressor. If the function must be simple, a simple learning algorithm, with a low variance and high bias, can help the machine learn. However, on many occasions, the function will be complex if there are many input variables or factors. The machine must act differently for different parts of the input vector. In such cases, the machine can only learn through a large training dataset. These cases also require the algorithms used to be flexible with low bias and high variance. Therefore, efficient learning algorithms automatically arrive at a tradeoff for the bias and variance depending on the complexity of the function and the amount of training data required.

Dimensionality of the input space

Dimensionality of the vector space is another issue that the engineer must deal with. If the input vector includes many features, the learning problem will become difficult even if the function only considers a few of these features as valuable inputs. Since there are many input variables, the input vector will have many dimensions, which can lead to confusion. This situation ultimately leads to the first issue that we discussed. So, when the input dimensions are large, the engineer should make an adjustment to the classifier to offset the effects of low variance and high bias. In practice, the engineer could manually remove the irrelevant features to improve the accuracy and efficiency of the learning algorithm. However, this might not always be a practical solution. In recent times, some algorithms can remove unnecessary features and retaining only the relevant ones. This concept is known as dimensionality reduction, which helps in mapping input data into lower dimensions to improve the performance of the learning algorithm.

Noise in the output values

The final issue on this list is concerned with the interference of noise in the desired output values. The output values are wrong in some situations since there is some noise associated with the sensors. There is a possibility that these values are wrong because of human error. In such cases, the learning algorithm should not look to match the training inputs with their exact outputs. For such cases, algorithms with high bias and low variance are desirable.

Other factors to consider

- The engineer must always consider the heterogeneity of data. She must choose an algorithm that considers the level of heterogeneity of the data. Some algorithms work better on datasets with a limit on the number of inputs used. Some examples are support vector machines, logistic regression, neural networks, linear regression and nearest neighbor methods. Nearest neighbor methods and support vector machines with Gaussian kernels work especially better with inputs limited to small ranges. On the other hand, there exist algorithms like decision trees that work very well with heterogeneous data sets.

- There is a possibility that the dataset is redundant. A few algorithms perform poorly in the presence of high redundancy.

This happens due to numerical instabilities. Examples of these types of algorithms are logistic regression, linear regression and distance-based methods. The engineer must include regularization to ensure that the algorithm performs better.

- While choosing algorithms, engineers need to consider the amount of non - linearities in the inputs and the interactions within different features of the input vector. If there is little to no interaction and each feature contributes independently to the output, algorithms based on distance functions and linear functions perform very efficiently. However, when there are some interactions within the input features, it is best to use the decision trees and neural network algorithms. These algorithms detect the interactions between the input vectors. If the engineer decides to use linear algorithms, he must specify the interactions that exist.

The engineer can compare various algorithms before she chooses the one to address a specific application. However, she must spend some time and collect training data and tune the algorithm to ensure it works for the application. If provided with many resources, it is advisable to spend more time collecting data than spending time on tuning the algorithm because the latter is extremely tedious. The most commonly used learning algorithms are neural networks, nearest neighbor algorithms, linear and logistic regressions, support vector machines and decision trees.

Chapter Four

Top Six Real Life Applications
of Machine Learning

We use many applications regularly that involve machine learning.

Image Recognition

The most common application of machine learning is image recognition. Most laptops and phones use the algorithm for image recognition. There are many situations when you can classify a certain object as an image. The measurements of every digital image always give the user an idea about the output of each pixel in the image.

If you were to look at a black and white image, the intensity of every pixel in the image serves as a measurement. If the image has M*M pixels, the measurement is M^2.

The machine splits the pixels into three measurements that give the intensity of the three primary colors (RBG). So, if there was an M*M image, there will be three M^2 measurements.

Face Detection

The most common category is the presence of a face versus the presence of no face. There can also be a separate category for every person in a database with multiple individuals.

Character Recognition

You can segment pieces of writing into images of small sizes where each image contains one character. These categories may comprise the 26 letters of the English alphabet, the first ten numbers and some special characters.

Speech Recognition

Speech recognition application is the translation of spoken words into actual text. Experts refer to it as Automatic Speech Recognition (ASR), Speech to text (STT) or Computer Speech Recognition (CSR).

The programmer uses spoken words to train the machine to recognize speech and convert the words into text. Facebook and Google are both using this method to train their machines. Machines use measurements to represent the signal of speech. These signals are further split into distinct words and phonemes. The algorithm uses different energies to represent the speech signals.

The details of the representation of signals are outside the scope of this book, but it is important to know that real values represent these signals. Applications on speech recognition often include voice user interfaces. These interfaces are call routing, voice dialing and other similar applications. These applications can also use data entry and other simple methods of processing information.

Prediction

Let us assume that a bank is trying to calculate the probability of a loan applicant defaulting on a repayment. To calculate this probability, the system will first have to identify, clean and classify the data that is available in groups. Analysts classify the data based on certain criteria. Prediction is one of the sought-after machine learning algorithms. If you were to look at a retailer, you can get reports on the sales that happened in the past. Now, you can predict what the sales may be soon. This will help the business make the right decision in the future.

Medical Diagnosis

Machine learning (ML) provides techniques; tools and methods that help a doctor solve prognostic and diagnostic problems in many medical domains. Doctors and patients can both use these techniques to enhance their medical knowledge and analyze the symptoms to obtain the prognosis. The result of this analysis will help to enhance the medical knowledge most doctors have. Doctors can also use machine learning to identify the irregularities in unstructured data, the interpretation of continuous data and to monitor results efficiently.

The successful use of different methods helps to integrate computer-based systems with the healthcare environment thereby providing the medical world with opportunities to enhance and improve treatments.

In medical diagnosis, the interest is to establish the existence of a disease and then identify the disease accurately. There are different categories for each disease that are under consideration and one category where the disease may not be present. Machine learning helps to improve the accuracy of a diagnosis and analyzes the data of the patients. The measurements used are the results of the many medical tests conducted on the patient. The doctors identify the disease using these measurements.

Statistical Arbitrage

Statistical Arbitrage, a term often used in finance, refers to the science of using trading strategies to identify the short-term securities one can invest in. In these strategies, the user can implement an algorithm on an array of securities based on the general economic variables and historical correlation of the data. The measurements used help to resolve the classification and estimation problems. The assumption is that the stock price will always lie close to a historical average.

Index arbitrage is a strategy that uses machine-learning methods. The linear regression and support vector regression algorithms help a user to calculate the different prices of funds and stocks. Using the principal component analysis, the algorithm breaks the data into dimensions to identify the trading signals as a mean reverting process.

Buy, hold, sold, put, call or do nothing are some categories under which the algorithm places these securities under. It then calculates the expected return for each security for the future. These estimates help the user decide which security to buy or sell securities.

Learning Associations

Learning association is the process of developing an insight into the association between groups of products. There are several products that reveal an association with one another although they seem unrelated. The algorithm uses the buying habits of customers to establish an association.

Basket learning analysis, which deals with studying about the association between products purchased by different customers, is an application of machine learning. If we assume that Amy has bought a product X, we can try to identify if she will buy product Y based on an association between the two products. Let us use the example of fish and chips to understand this concept. If a new product enters the market, the association between the existing products will change. If one knows these relationships, they can identify the right products to give their customers. To increase purchasing power, a company can choose to introduce products in pairs.

Big Data analysts use machine-learning algorithms to identify if there is a relationship between different products. Algorithms use probabilities to establish a relationship between products.

Chapter FIve

Machine Learning Algorithms

The many subjects that have laid the foundation of machine learning are discussed in the first chapter. This chapter covers some of the machine learning algorithms that engineers use to train the machine.

Dimension Reduction Methods

There are millions or billions of records and variables in the databases that engineers use to derive the training dataset. It is impossible to conclude that these variables are not dependent on one another with absolutely any correlation between them. It is important for the engineer to remember that there are multiple collinearities that exist between the variables. In this situation, the predictor variables are correlated in some way that can affect the output.

A lot of instability arises in the solution set when there is multicollinearity between variables leading to incoherent results. For instance, if you look at multiple regressions, there are multiple correlations between the predictor variables that have a significant impact on the output set. However, individual predictor variables may not have a significant impact on the solution set. Even if the engineer identifies a way to remove such instability, there are times when the users may include variables with a high level of correlation between them. In this case, the algorithm will focus on some parts of the input vector more than the others.

When the dataset has many predictor variables, an unnecessary complication arises where the algorithm must identify a model between the predictor and response variables. This situation complicates the analysis and its interpretation and violates the principle of parsimony. The principle of parsimony states that an analyst should always stick to a certain number of predictor variables, which makes it easy for human beings and now machines to interpret the results. If the engineer retains too many variables, there is a possibility that there could be over fitting, which will hinder the analysis. The new dataset that the engineer uses will not behave in the same way as the training dataset or the predictor data.

There is also the question of how the analysis performed at the variable level will miss the relationships that lie between the predictors. For instance, there can be numerous single predictor variables that fall into a single group or component that will address only one aspect of the data. If you look at a person's account, you will need to group the account balance and any deposits or savings made from that account into one category alone.

There are certain applications, such as image analysis, where the full dimensionality of the variable is retained which makes the problem intractable. For instance, a face classification system based on 256×256-pixel images could potentially require vectors of dimension 65,536. Human beings can discern and understand certain patterns in an image at a glance. These patterns can elude the human eye if the output represents them as an algebraic equation or graph. However, the most advanced visualization techniques also do not go beyond five dimensions. How do you think we can identify the relationships that could exist between a massive data set that has thousands of variables?

The goal of dimension reduction methods is to use the structure of correlation among the different predictor variables to accomplish the following goals:

- Reduce the number of predictor components in the data set

- Ensure that these predictor components are independent of one another

- Provide a dynamic framework which will help in the interpretation of the analysis

The most common dimension reduction methods are Principal Component Analysis (PCA), User Defined Composites and Factor Analysis.

Clustering

Clustering is a machine learning technique that groups different data points in the dataset that are similar. An engineer often uses this algorithm to classify and categorize the data points into specific groups. Data points or variables in the same group should bear some similarities, while variables in different groups should bear no similarities. Clustering is an unsupervised machine-learning algorithm that an engineer commonly uses for statistical data analysis.

Data scientists use clustering analysis to derive insights from the data by identifying the groups that the data points or variables fall into when the algorithm runs. This section covers some clustering algorithms that an engineer uses to train the machine.

K Means

The K-Means clustering algorithm a well-known clustering algorithm. This is a concept that every engineer and data scientist must know the algorithm well since it is easy to implement in a code.

- The algorithm must first select the number of groups or classes and initialize the center points of these classes. If you are unsure of how many classes to use, look at the data and identify the different categories and classes within the data. The center points of these classes are the lengths of the vectors.

- The algorithm classifies the data points by calculating the distance between the point and all the center points. The engineer then uses this distance to categorize that data point in the class whose center is the closest to that point.

- Using the classified points, the algorithm computes the center of the data points or variables in the class using the mean.

- The engineer must repeat these steps multiple times until the center of the groups does not change between the iterations. You can also initialize centers in the groups at random and then select an iteration that represents the best results.

This clustering algorithm has an advantage that it is simple and fast to use since you are only computing the distance between the variables and the center of the groups. On the other hand, it also has a few disadvantages. You must select the number of classes by looking at the data and this is not an ideal method to use since you want some insight from the dataset. There is also the point where you should select the center points of groups at random, which can lead to different results in different iterations. Thus, you may not obtain a consistent output that makes it difficult to select the center points of the group.

Mean Shift Clustering

An engineer uses the mean shift-clustering algorithm to find dense areas in the dataset. This algorithm also uses the center points of every group. However, the goal of the algorithm is to update the possible center points of the class within the sliding-window to locate the actual center point. The algorithm removes the center points after the processing stage to remove duplicates and form the final set of cluster points and place them in their groups.

- Let us consider a set of points in a two-dimensional space. The first step is to define the point around which the circular sliding window is positioned. The circular window should have a radius r called the kernel. This algorithm is a hill-climbing algorithm and it constantly moves the kernel to denser regions until the values converge.

- The sliding window moves to denser regions at every iteration. The algorithm does this by shifting the center point of the group to the mean point. The density of the points in the sliding window is directly proportional to the number of data points in it. Therefore, when the algorithm shifts the mean of the points in the window, it moves towards areas in the data with a higher density.

- The sliding window moves according to the change in the mean and there is no specific direction in which the window moves.

- The algorithm repeats the first three steps using different sliding windows until it can categorize all the data points in the set into different sliding windows.

In this method, you do not have to select the number of classes or clusters, which is a great advantage. It is also good that the center points converge to the mean of the highly dense data region. This algorithm understands data and fits any application that is data-driven. The selection of the kernel is trivial, which is the only drawback of this method.

It is best to use the K-Means algorithm if you are a beginner. Once you understand this algorithm, you can use other clustering algorithms in your analysis.

Regression Modeling

Regression modeling is an algorithm that an engineer uses frequently to estimate the values of continuous target variables. There are many regression models that you can choose from. However, the simplest form of the regression algorithm is the simple linear regression model. In this model, the algorithm defines the relationship between a continuous predictor variable and continuous response variable using a straight line. There are models that use multiple predictor variables to define the response variable.

Apart from the models mentioned above, engineers use two algorithms called the least squared regression and logistic regression methods to train machines. However, the assumptions of the regression model have created a disparity. It is imperative that the engineer validates these assumptions before writing the algorithm and building the model. If the engineer builds a model and uses it without verifying the assumptions, there is a possibility that the engineer cannot use the output since the model may have failed without the knowledge of the engineer.

When the engineer obtains the result, she must ensure that there is no linear relationship between the variables of the model. There are times when the dataset has variables that have a hidden linear relationship. However, there is a systematic approach, called inference, which the engineer can use to determine the linear relationship. The inference methods that an engineer can use to determine the relationship are:

- The t-test for the relationship between the response variable and the predictor variable

- The confidence interval for the slope, $\beta 1$

- The confidence interval for the mean of the response variable given a value of the predictor

- The prediction interval for a random value of the response variable given a value of the predictor

The inferential methods described above depend on the assumptions that the engineer makes at the beginning of the process. It is easy for the engineer to assess whether the data adheres to the assumptions using two graphical methods - a plot of the normal probability and a plot of the standardized residuals against the predicted or fitted values.

A normal probability plot is a quantile-quantile plot of the quantiles of a distribution against the quantiles of the standard normal distribution for the purposes of determining whether the specified distribution deviates from normality. In a normality plot, the engineer makes a comparison between the values observed for the distribution of interest and the expected number of values from the normal distribution. If the distribution is normal, the bulk of the points in the plot should fall on a straight line; systematic deviations from linearity in this plot indicate non-normality. An engineer can evaluate the validity of the regression assumptions by verifying if some patterns exist in the residual plot. If there is a pattern, the engineer must identify which assumption does not hold true. However, if no patterns exist, the assumptions remain intact.

If these graphs indicate violations of the assumptions, we may apply a transformation to the response variable y, such as the ln (natural log, log to the base e) transformation. If the relationship between the response variables and the predictor variables is not linear, the algorithm can use transformations. We may use either "Mosteller and Tukey's ladder of re-expression" or a "Box-Cox transformation."

Regression with Categorical Predictors

Thus far, our predictors have all been continuous. However, the engineer can choose to give categorical predictor variables as inputs to regression models using dummy or indicator variables.

For use in regression, the algorithm transforms categorical variables with k categories to a set of (k-1) indicator variables. An indicator variable is a binary 0/1 variable, which takes the value 1 if the observation belongs to the given category and takes the value 0 otherwise.

Logistic Regression

The engineer uses the logistic regression algorithm to approximate the relationship between a set of continuous predictor variables and a categorical response variable. It is difficult to use linear regression in such instances. The logistic regression algorithm is like the linear regression algorithm except that the former uses methods to estimate the relationship between the predictor and response variables. In other words, the engineer uses the linear regression method to approximate the relationship between a set of continuous predictor variables and a continuous response variable. Logistic regression, on the other hand, refers to methods for describing the relationship between a categorical response variable and a set of predictor variables.

An attractive attribute of the linear regression algorithm is that the engineer can obtain closed-form solutions for the optimal values of the coefficients through the least-squares method. However, in logistic regression, the engineer must use the likelihood functions to estimate the parameters of the model. This method obtains parameters by maximizing the likelihood of the observing data.

The engineer can find the value of the maximum likelihood estimators by differentiating the likelihood function, $L(\beta|x)$, with respect to each parameter and then setting the resulting forms to be equal to zero. Unfortunately, unlike linear regression, closed-form solutions for these differentiations are not available. Therefore, the engineer must apply other methods like iterative weighted least squares.

Logistic regression assumes that the relationship between the predictor and the response is nonlinear. In linear regression, the response variable is a random variable $Y = \beta_0 + \beta_1 x + \varepsilon$ with conditional mean $\pi(x) = E(Y|x) = \beta_0 + \beta_1 x$. The conditional mean for logistic regression takes on a different form from that of linear regression.

Variable Selection Methods

Different variable selection methods are available for the engineer to use to determine which variable must be a part of the dataset, including forward selection, backward elimination, stepwise selection and best subsets. These variable selection methods are essentially algorithms that construct the model with the optimal set of predictors.

Forward Selection Procedure

The forward selection procedure does not use any variables.

Step 1: For the first variable to enter the model, select the predictor most highly correlated with the target. (Without loss of generality, denote this variable x1.) If the resulting model is not significant, stop and report that no variables are significant predictors; otherwise, proceed to step 2.

Step 2: For each remaining variable, compute the sequential F-statistic for that variable given the variables already in the model. For example, in this first pass through the algorithm, these sequential F-statistics will be $F(x2|x1)$, $F(x3|x1)$ and $F(x4|x1)$. On the second pass through the algorithm, these might be $F(x3|x1, x2)$ and $F(x4|x1, x2)$. Select the variable with the largest sequential F-statistic.

Step 3: For the variable selected in step 2, test for the significance of the sequential F-statistic. If the resulting model is not significant, stop and report the current model without adding the variable from step 2. Otherwise, add the variable from step 2 into the model and return to step 2.

Backward Elimination Procedure

The backward elimination procedure begins with all the variables or all a user-specified set of variables in the model.

Step 1: Perform the regression on the full model; that is using all available variables. For example, perhaps the full model has four variables, namely, x1, x2, x3 and x4.

Step 2: For each variable in the current model, compute the partial F-statistic. In the first pass through the algorithm, there will be $F(x1|x2, x3, x4)$, $F(x2|x1, x3, x4)$, $F(x3|x1, x2, x4)$ and $F(x4|x1, x2, x3)$.

Select the variable with the smallest partial F-statistic. Denote this value as F-min.

Step 3: Test for the significance of F-min. If F-min is not significant, remove the variable associated with F-min from the model and return to step 2. If F-min is significant, stop the algorithm and report the current model. If this is the first pass through the algorithm, the current model is the full model. The algorithm reduces the number of variables if this is not the first pass.

Stepwise Procedure

The stepwise procedure represents a modification of the forward selection procedure. A variable that is present in the forward selection algorithm may become insignificant once other variables are present in the model. The stepwise procedure checks on this possibility by performing at each step a partial F-test using the partial sum of squares for each variable currently in the model. If there is a variable in the model that is no longer significant, the algorithm eliminates the variable with the smallest partial F-statistic from the model. The procedure terminates when the algorithm cannot enter or remove any variable.

Best Subsets Procedure

For data sets where the number of predictors is not too large, the best subsets procedure represents an attractive variable selection method. However, if there are more than 30 or so predictors, the best subsets method encounters a combinatorial explosion and becomes intractably slow. The best subsets procedure works as follows:

Step 1: The analyst specifies how many (k) models of each size he or she will like to report, as well as the maximum number of predictors (p) the analyst wants in the model.

Step 2: In this step, the algorithm builds all models of one predictor: for example, $y = \beta_0 + \beta_1$ (sugars) $+\varepsilon$, $y = \beta_0 + \beta_2$ (fiber) $+\varepsilon$ and so on.

The algorithm calculates their R2, R2adj, Mallows' Cp and reports the best k models based on these measures.

Step 3: The algorithm builds all models of the two predictor variables: for example, $y = \beta_0 + \beta_1 (sugars) + \beta_2 (fiber) + \varepsilon$, $y = \beta_0 + \beta_1 (sugars) + \beta_4 (shelf2) + \varepsilon$ and so on. The algorithm calculates their R2, R2adj, Mallows' Cp and reports the best k models based on these measures.

The procedure continues in this way until the algorithm reaches the maximum number of predictors (p). The analyst then has a listing of the best models of each size 1, 2... etc. to assist in the selection of the best overall model.

Chapter Six

Machine learning Projects

In this chapter, we will cover some simple and easy machine learning projects that a beginner can use. It is always good to work on projects since they are a great investment of your time and make learning fun. You will notice that you are making faster progress when you work on the projects.

You can read volumes of theory, but that can never help you build confidence in the subject when compared to hands-on practice. Most people believe that they become masters when they read textbooks and articles. But, when they try to apply the theory, they notice that it is harder than it looks.

When you work on projects, you improve your skills and have the chance to explore interesting topics. You can also add these projects to your portfolio, which will make it easier for you to land a job. You can complete the projects in this chapter over the weekend. However, you can expand them to build longer projects if you want.

Data Analysis

The data analysis project is one that most people begin with since it helps to hone some statistical skills. By working on this project, you can build your practical intuition around machine learning. The goal of this exercise is to help you use different models and apply those models to a variety of datasets. There are three reasons why you should begin with this project.

- You will learn to identify which model will fit your problem. For example, some datasets may have missing information. When you work on this project, you will know which model you should use for such datasets. You can always dig through texts and articles to learn which model is better, but it is better to see it in action.

- You will develop the art of creating prototypes of models. It is often difficult to identify if a model will work best for the problem without trying it.

- You must keep the process flow or workflow of the machine in mind when you build a model. You will master this technique when you work on this project. For example, you can practice:

  o How to import data

  o Which tools should you use to clean data

  o How to split the data into training sets

  o Pre-processing of data

  o What transformations must you make to the data

Since you use different models, you can focus on the development of the skills mentioned above. For instructions, you can look at the Python (sklearn) and R (caret) documentation pages. You must also practice classification, clustering and regression algorithms.

Social Media

Social media is synonymous with big data because of the volumes of content that users generate. You can mine this data and keep tabs on trends, public sentiments and opinions. Facebook, YouTube, WhatsApp, Instagram and many other platforms have data that you can use to achieve your objective.

Social media data is important for branding, marketing and for a business since every generation spends more time on social media when compared to its preceding generation. Twitter is one of the platforms you should start with when you begin to practice machine learning. You have interesting data sets and metadata that can open an endless path for analysis.

Fantasy Leagues or Sports Betting

If you have read the book Moneyball, you know how the Oakland A's changed the face of baseball through their analytical scouting. Based on the analysis, the team built a highly competitive squad and bought the players for almost a third of the price when compared to the Yankees players. If you have not read the book yet, you must grab a copy and read it now.

There are large volumes of data in the sports world that you can play with. You can use data for games, teams, players and scores to analyze patterns. These data are available online. For instance, you can try the following projects and see where they lead you.

- You can bet on specific box scores in different games based on the data available on the Internet right before a game begins. You can analyze the data and identify where you should place your bets to increase your chances of winning.

- You can scout for talent in schools and colleges using the statistics for different players. This analysis can help college scouts identify the players they want on their team.

- You can become a manager for a fantasy team and use your analysis to win games and leagues.

The sports domain is especially useful since it helps a beginner practice exploratory analysis and data visualization. These skills will help you identify what types of data you should include in your analysis.

Stock Market

An aspiring data scientist will always base his projects on the stock market since that is like Disneyland for them. The stock market is an interesting area to work on since you have different types of data available. You can find data on fundamentals, prices, stocks, economic scenarios, macroeconomic and microeconomic indicators, volatilities and many more.

The data available is granular which makes it easier for people to obtain historical data, also called time series data, for every company. Using this data, you can build creative trading strategies. Since financial markets have small cycles, you can validate any prediction you make immediately.

- You can analyze the data and assess the movement of prices for different stocks for a period of 6 months. You should use the quarterly reports provided by the company to make this prediction.

- You can use a time series models and recurrent neural network models to forecast the correlation between the stock prices. You can also assess the volatility of the prices.

It is simple to build a financial trading model to improve your skills in machine learning. However, it is difficult to profit from these models here. Experts recommend that you do not use real money to trade until you have perfected the model.

Writing Machine Learning Algorithms

It is important for a beginner to learn how to write machine-learning algorithms from scratch for two reasons:

- There is no better way to understand the mechanics of the algorithms since you must think about the process flow of the algorithm thereby helping you master algorithms.

- You can translate statistical and mathematical instructions into working code, which allows you to apply algorithms in your academic research.

It is best to start off with simpler algorithms since you must make hundreds of decisions. Once you are comfortable with simple algorithms, you can extend the functionality of those algorithms to complex algorithms like regression and clustering.

Your packages may be slow and may not be as fancy as the algorithms in existing packages. It took the developers years to build these packages.

Chapter Seven

Artificial Intelligence

An amazing development has taken place over the last few years. You may have watched how robots in Star Wars were able to perform so many actions and maybe, just maybe, you would wondered how wonderful it would be if it were to happen in this decade. You may not have seen it coming, but this was an inevitable turn of events – the emergence of Artificial Intelligence (AI). Everywhere we look today; we come across some intelligent systems that talk to us – Siri, Google Assistant – offering us advice and offering us recommendations. These systems improve almost every year to improve its interpretation of images, voice recognition and to drive cars based on different techniques used by Facebook and Google's Deep Learning Efforts. Other work always aims to understand and generate machines that can understand our language and communicate with us.

The reemergence of AI has caused a lot of confusion since there are so many companies that have begun to explore the scene. How do we make sense of any of it?

Let us start with a simple definition of AI. Artificial Intelligence or AI is a field of computer science that aims to develop computers that can perform tasks that human beings can perform.

AI has had some excellent runs. In the early sixties, experts made promises about the abilities of machines and what they can do. In the eighties, artificial intelligence revolutionized businesses. But in those eras, the promises made were too difficult to deliver. So, what makes the latest developments in AI any different? What makes the systems developed now any different from the diagnostic programs and neural nets of the past? There are some reasons why the developments in this era are different from the last.

Increased Computational Resources

The computers we have in this era are faster and can think harder thereby increasing the computational power. The techniques used earlier worked well only in the past, but now there is a necessity to improve the computational grid and expand it.

Deeper Focus

AI has shifted away from looking at smaller aspects of data to look at specific problems. The systems now are capable of thinking about a problem as opposed to daydreaming without any problem. Systems like Cortana and Siri work well within limited domains that focus on speech and image recognition (covered in chapter four). These platforms use words and sentences to understand what you need and provide you with the required output.

Alternative Reasoning Models

Experts use alternative reasoning models, which assume that the systems do not have to reason like human beings to be smart. The machine can think like a machine.

It is these factors put together that have given the world the first renaissance of intelligent machines that have become a part of our lives. We are now using different machines to ease our work and lives.

Knowledge Engineering

Different types of learning study the problem and issues with knowledge engineering. The systems developed these days use their ways to learn. The bottleneck in the systems in the past helped us add more rules to avoid such bottlenecks in the further processing of data. Most approaches in the modern times focus on learning these rules automatically.

Growth of Data

Over the years, the data collected has increased by a vast amount and the engineer feeds the machine with this data to enhance the machine's computational abilities. This goes to say that learning systems get better at understanding more data and would now be able to look at thousand examples as opposed to only a few hundred.

Exploring AI

As mentioned earlier, some engineers allow machines to think like machines. The key word here is most. Some developers still look to create machines that are capable of thinking like human beings. The only opposition to this is the doubt about the intentions of the machine. Some people are worried about what machines could be capable of if they could think exactly like human beings.

There are two types of artificial intelligence – strong AI and weak AI.

Strong AI

The sphere of strong AI deals with the development of machines and systems that are capable of thinking like humans. This entails that the machines should be able to reason out and solve problems the way humans do. They should be capable of explaining their solutions as well. Basically, these machines need to have abilities like those of human cognition.

Weak AI

As mentioned earlier, people have qualms about designing machines capable of human thinking. So, weak AI deals with the development of machines that work but not at the level of human thinking. Essentially, it deals with the creation of systems and machines that can behave like humans and perform tasks that we can but are devoid of any thinking capabilities that rival human cognition. A great example of such a machine is the Deep Blue made by IBM. The machine became a brilliant chess player, but it was unable to explain its moves or show any cognitive abilities.

Anything in Between

In addition to the two types of AI mentioned above, engineers and scientists are developing machines that are in between the above types. These types of systems are capable of learning and can also understand the human reasoning. However, they are not subservient to it. The trend in recent times has shifted to these types of machines. The essence of this type of development is to use human cognitive skills and emotions to train the machines but not use them as the models for the machines.

The main conclusion of the previous few paragraphs is that, as opposed to the consensus from previous times, machines do not have to think like humans to be smart. However, a machine that thinks like a machine can learn using training datasets.

Assessing Data using AI and Machine Learning

All the major consumer systems around the world are constantly trying to estimate human beings. For instance, online shopping websites such as eBay and Amazon are constantly using machine learning and AI to figure out your likes and dislikes so that they can put forth a list of recommendations for you. As mentioned earlier, the machines learn from the data that you provide them with when you browse and look for products (transactional information). Even web pages like Facebook use this concept to recommend pages that you might want to follow.

Using profile data is only a small step in this process. The machines also use information from different categories for training purposes. Most machines also recommend some products using information like your budget, your likes and your buying history. These systems also look at the customers who live around you to further refine the list of recommendations it presents you with.

The results of these processes are almost always a group of characteristics:

For instance, if you were someone who just purchased a garden and is looking to grow plants on it, you would probably go online to look for gardening tools. The next time that you would visit that website, you would see recommendations related to gardening. This is how the machine develops that list - it cross-references the products you looked for with products that people have already bought and then looks at what else they purchased. It accesses this list and presents it to you as a recommendations list.

Chapter Eight

Glossary

This chapter provides a list of words and their definitions when we talk about machine learning.

Data Science

Data science, in its truest form, represents the resource and process optimization of the data analysis. Through data science, you can produce data insights and use those insights to improve your investments, business, health, lifestyle and social life. For any pursuit or goal, you can use data science methods to help you understand and predict the path you must take to meet your goals and objectives. You will also be able to anticipate any obstacles or hurdles that come your way.

Data Mining

Many businesses use the process of data mining to convert the collected raw data into information that it can use. There are specialized tools that the business can use to detect patterns in large-scale information. These patterns help the business learn more about its consumers and respond to their concerns while developing strategies that will help to increase your revenue. The objective of data mining is to increase profits. Data mining is effective only when the process of collecting and storing data is done the right way.

Artificial Intelligence

Artificial Intelligence or AI is a field of computer science that aims at developing computers that can perform tasks that human beings perform.

Additive Property

The next axiom that is important to note can only be true when both events A and B are mutually exclusive.

$$P(A+B) = P(A) + P(B)$$

This axiom states that the probability of both events A and B occurring is the same as the sum of the individual probabilities of the events occurring if and only if both events exclude each other. For example, if A is the event that we get a six on rolling a die and B is the event that we get a five rolling a die, then this axiom hold. But, if B is the event where we get an even number, this set will include the number six making the axiom false.

Regression

Regression is closely related to classification. Classification is directly concerned with the prediction of discrete classes. When the class has continuous variables, the engineer uses machine learning to obtain the output. Linear regression is an example of regression techniques. To learn more about regression, please read the previous chapter.

Joint Probability

This property can be expressed as follows:

$$P(a, b) = P(A=a, B=b)$$

This property can be read as the probability of a and b is the same as the probability that event A turns out in state 'a' and event B turns out in state 'b.'

Bayes' Rule

This rule is used to identify the conditional probability when P (A, B) is not known. The equation used is as follows:

$$P(A|B) = [P(B|A)*P(A)]/P(B)$$

These various axioms are a great way to understand the logic behind Markov Models. Let us now look at the mathematical aspects of the Markov Model. As mentioned above, Markov Models is a concept that was discovered in the year 1916 by Andreevich Markov, a scientist who was studying and analyzing the frequency of different types of words in Pushkin's poems. These models have now become an integral model to use while working with data science, artificial intelligence and machine learning.

Classification

Classification is concerned with separating data into unique classes using models. The engineer builds these models using a training dataset that is named to help the algorithm learn. Users input real-time data with the classes that are present in the model. This will help the model predict the relationship that exists within the data based on what the model has learned from the training dataset. Well – known classification schemes are support vector machines and decision trees. Since these algorithms will need an explicit definition of classes, classification is a form of supervised machine learning.

Support Vector Machines

Support Vector Machines (SVMs) allow the user to classify linear and nonlinear data. They work by transforming the training dataset into higher dimensions. The model inspects these dimensions to calculate the optimal boundary separation between classes. In SVMs, these boundaries are called hyperplanes. The algorithm identifies hyperplanes using support vectors or the instances that define the classes and their margins, which are lines that are parallel to the hyperplanes. These are defined by the shortest distance between the hyperplane and the support vector associated with it.

The goal behind using an SVM is to identify the hyperplane that separates two classes if there are many dimensions. This process helps to delineate the member classes in the dataset. When the algorithm repeats the process several times, it generates enough hyperplanes that can help to separate dimensions in an n – dimension space.

Clustering

Clustering is a technique that an engineer or data scientist uses to analyze data that does not include pre-labeled classes or any class attribute at all. The instances in the data are grouped using the concept of maximizing the similarity within classes and minimize the similarity between classes. This loosely translates into the clustering algorithm that identifies and groups the instances that are like each other. The best – known clustering algorithm is k – means clustering. Clustering does not require the pre-labeling of instance classes; therefore, it is a form of unsupervised machine learning meaning that the algorithm learns more from observation as opposed to learning by example.

Association

Association can be explained easily by introducing a market basket analysis, which is a task that it is well–known for. This type of analysis always tries to identify the association that exists between different data instances that are chosen by any shopper and placed in their basket. This could either be real or virtual and the algorithm always assigns confidence and support measures for comparison. The value of this always lies in customer behavior analysis and cross marketing. Association algorithms are generalizations of market-based analyses and are like classification algorithms in the sense that any attribute can be predicted when using association. Apriori is one the best – known association algorithms. If you have deduced that association is an example of unsupervised machine learning, then you are right.

Machine Learning

Although this book covers machine learning in detail, we will just look at it one more time. Machine learning is concerned with how a computer can be constructed to automatically improve the experience of the user. Machine learning is an interdisciplinary science that employs techniques from different fields like computer science, artificial intelligence, mathematics and statistics and so on. The main aspects of machine learning research include algorithms that help to facilitate this improvement from experience. Artificial intelligence, data mining and computer vision are some fields that use these algorithms.

Decision Trees

Decision trees are recursive, divide – and – conquer and top-down classifiers. These trees are composed of two main tasks: tree pruning and tree induction. The latter is the task where a set of pre-classified instances are taken as inputs after which decisions are made based on which attributes are split on thereby splitting the dataset and recursing on the resulting split datasets until every training instance is categorized. While building the tree, the main goal is to split all the attributes to create the child nodes that are pure. This ensures that the number of splits needed to classify the instances in the dataset is small. The purity of the child nodes is measured using information that relates to the volume of information that the machine must know about a previous instance and how it should categorize it.

A complete decision tree model can always be complicated and may contain some unnecessary structure, which make it difficult for human beings to interpret the output. Unnecessary structure or branches are removed through tree pruning, which makes the decision tree easily readable, more efficient and accurate for human beings to comprehend. This increased accuracy due to tree pruning helps to reduce over fitting.

Fundamental Axioms

Let us look at the various axioms that are used in machine learning to understand the math that supports the model. One of the most fundamental axioms can be expressed as follows:

$$0 < P(A) < 1$$

This states that the probability of any event occurring is always going to be greater than zero but less than one, both inclusive. This implies that the probability of the occurrence of any event can never be negative. This makes sense since the probabilities can never be more certain than a hundred percent and least certain than zero percent.

Deep Learning

Deep learning is a method that uses neural networks to identify solutions. This type of learning uses different layers and nodes of input that send signals to the hidden layers in the network to identify the solution to any input. The work in deep learning is defined by how the human mind learns and how the mind learns. It also considers how calculations and computations take place in the cerebral cortex of the human brain.

Generative Model

In statistics and probability, a generative model is used to generate data sets when some parameters are hidden. These models are used in machine learning to either model the data directly or used as an intermediate step to form a conditional probability density function.

Conclusion

The Industrial Revolution led to the development of machines. This development started small, where a machine was used to perform laborious tasks. Since then, machines have entered the lives of every human being and they are here to stay. This increased dependency on machines led to the development of artificial intelligence. Scientists and engineers are looking for different ways to train machines into performing difficult and intelligent tasks. Experts from different fields of study are now using machines to enhance the processes. Therefore, it is of utmost importance to understand what machine learning is and how it improves life.

This book provides information on what machine learning is and types of machine learning. It also gives you information on the subjects that laid the foundation for machine learning. You will gather information on different algorithms used in machine learning. As a beginner, it is always good to practice some algorithms used in machine learning to enhance your understanding. There are some projects in the book that you can complete over the weekend or extend them if you want to. It is important to practice as much as you can to improve your knowledge on machine learning. It is difficult to remember the many words that are used in machine learning. There is a glossary of words in the last chapter of the book that you can go through.

Thank you for purchasing the book. I hope you have gathered all the information necessary for machine learning.

Finally, if you enjoyed this book, then I'd like to ask you for a favor, would you be kind enough to leave a review for this book on Amazon? It'd be greatly appreciated!

Resources

https://www.kdnuggets.com/2016/08/10-algorithms-machine-learning-engineers.html

https://towardsdatascience.com/linear-regression-detailed-view-ea73175f6e86

http://www.statisticssolutions.com/what-is-linear-regression/

https://towardsdatascience.com/logistic-regression-detailed-overview-46c4da4303bc

https://www.statisticssolutions.com/what-is-logistic-regression/

https://medium.com/app-affairs/9-applications-of-machine-learning-from-day-to-day-life-112a47a429d0

https://machinelearningmastery.com/supervised-and-unsupervised-machine-learning-algorithms/

https://searchenterpriseai.techtarget.com/definition/supervised-learning

https://towardsdatascience.com/supervised-vs-unsupervised-learning-14f68e32ea8d

https://www.datarobot.com/wiki/unsupervised-machine-learning/

https://home.deib.polimi.it/matteucc/Clustering/tutorial_html/

https://towardsdatascience.com/the-5-clustering-algorithms-data-scientists-need-to-know-a36d136ef68

http://blog.aylien.com/10-machine-learning-terms-explained-in-simple/

https://www.forbes.com/sites/bernardmarr/2016/12/06/what-is-the-difference-between-artificial-intelligence-and-machine-learning/#2769c962742b

11593783R00035

Made in the USA
San Bernardino, CA
06 December 2018